inicia DUAL

SECONDARY

PLASTICS AND TEXTILES

W0211351

Oxford EDUCACIÓN

Contents

Oxford Projects

A digital task which allows you to investigate like a real scientist through the use of new technology. It includes activities, digital simulations, videos... all of which will help you understand the contents better.

The digital activity relates to the final task which you have to do at the end of each unit.

This is your book

PRESENTATION

In order to give you an overview of the unit, the presentation includes a full-page image and some stimulating questions to elicit knowledge which will help you relate the unit content to aspects of your daily life as well as showing you how to apply it to real life.

We recommend that you answer the questions as you advance within the unit.

The link by the picture will take you to a short introductory video.

FINAL TASK

There is also a final task which we will prepare you for as you go along the unit. It can be carried out individually or in groups.

It is identified with a task icon which marks the activities and sections related to it throughout the unit.

The same task is also available in digital format in **Oxford Projects.**

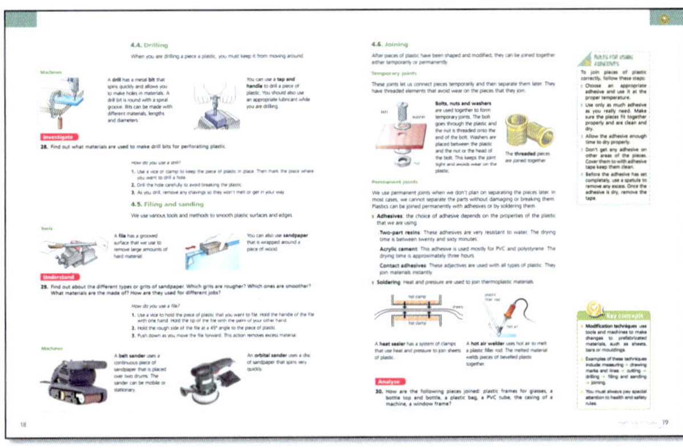

CONTENT DEVELOPMENT

On these pages the main contents you are going to study are explained. The activities are categorised according to Bloom's taxonomy, aimed at encouraging critical thinking skills.

In the margin of some pages you will find a brief summary of the content (*Key concepts*).

PROCEDURES

These pages let you discover interesting methods and techniques to work with the materials and tools you can find in your surroundings, applying the contents learnt in the unit.

ANALYSIS OF TECHNOLOGICAL PRODUCTS

These pages set out procedures and exercises on how to analyse a technological product related to the unit. The axis around which technology is articulated is the process of resolution of technological problems. In order to know and improve products, it is necessary to analyse them.

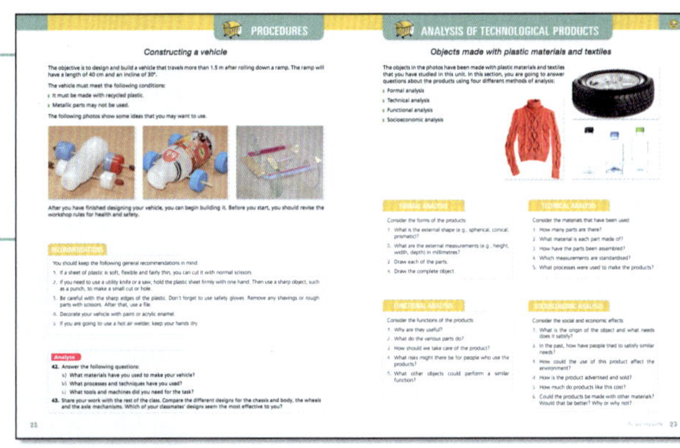

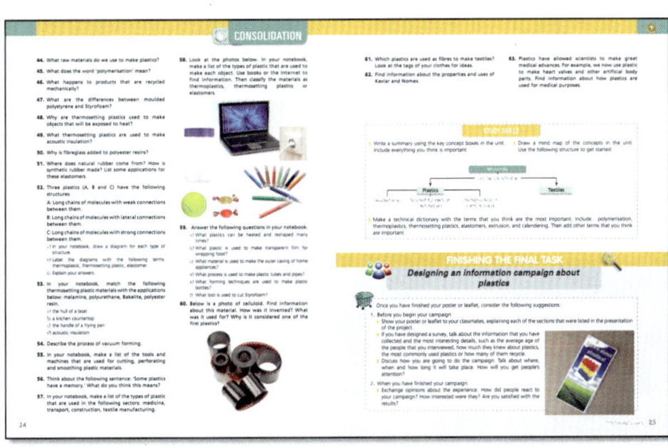

CONSOLIDATION

This double page of activities serves you to revise what you have learnt and extend some aspects seen. In the highlighted section *Study skills* you are encouraged to elaborate a summary, conceptual map and technical glossary.

FINISHING THE *FINAL TASK*

We provide you with guidance on how to carry out the task and present your results.

You have a **DUAL book**: a printed book and its electronic version, which includes resources for you to use together with the unit. In order to access the electronic version, use the code in the book and follow the instructions. You will be able to work either on line or offline.

Digital icon: this icon reminds you that, in the electronic version of you DUAL book you will find: *Oxford projects*, animations, videos, relevant web links and the interactive version of the course book activities.

PLASTICS AND TEXTILES

YOU WILL LEARN TO...

- Understand the properties and applications of plastics and how they are made.
- Classify plastic materials (thermoplastics, thermosetting plastics and elastomers) and identify their basic applications.
- Identify different types of plastic in the most typical technical applications.
- Understand industrial processes and apply techniques correctly while following appropriate health and safety rules.
- Evaluate the environmental impact of the exploitation, transformation and disposal of plastic materials, as well as the benefits of recycling.
- Understand the properties, classification and production of textiles.
- Identify different textiles by their applications.

- *What do you see on the beach in the photo above?*
- *What materials are those things probably made of?*
- *Why should people recycle these types of materials?*

FINAL TASK

Designing an information campaign about plastics

In this unit, you will design an information campaign about plastics. This campaign will include the design and distribution of posters and leaflets at your school and / or in your neighbourhood.

For your campaign, you should choose everyday products made of plastic. You should also keep in mind the points listed in the diagram.

You may also add other sections, such as:

- alternative materials that are less harmful.
- other products that perform the same functions.

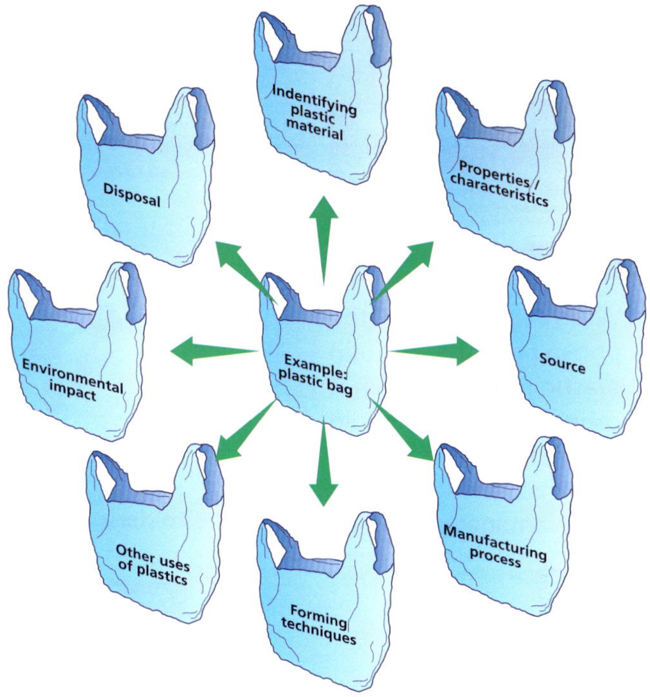

What do you have to do?

① Find, compare and select information.
② Collect information and create documents, using a word-processing program.
③ Design an informative poster or leaflet.
④ Display or distribute your work.

Characteristics of posters and pamphlets

- DIN A3 paper is typically used for posters. DIN A4 paper is used for leaflets.
- You should explain all of the points above.
- The information should be clearly organised and space should be used effectively.
- The content should be clear and concise.
- You should include charts and images, such as drawings or photos.

Before you begin... Answer

Think about your project. Then answer the questions below:

1. What do you know about the plastic objects that you use every day?
2. Can you identify different types of plastics? How are they typically used?
3. Where does plastic come from? What is it made of?
4. What are the environmental effects of plastics? Think about how we produce them and what we do with plastic waste.
5. In addition to the points in the diagram above, what other sections would you add to your poster or leaflet?
6. Write a survey about plastics with a maximum of 10 questions. Look for information about different types of surveys.

1. PLASTIC MATERIALS

Analyse

1. Which objects in the photo probably contain plastic materials? What other materials could be used? What are the advantages and disadvantages of plastic?

In general, plastic materials are a normal part of everyday life. They also play important roles in sectors such **packaging**[1], telecommunications, transport, construction, medicine, agriculture and information technology.

Plastics consist of long chains of **molecules**[2], mostly made of the element *carbon*.

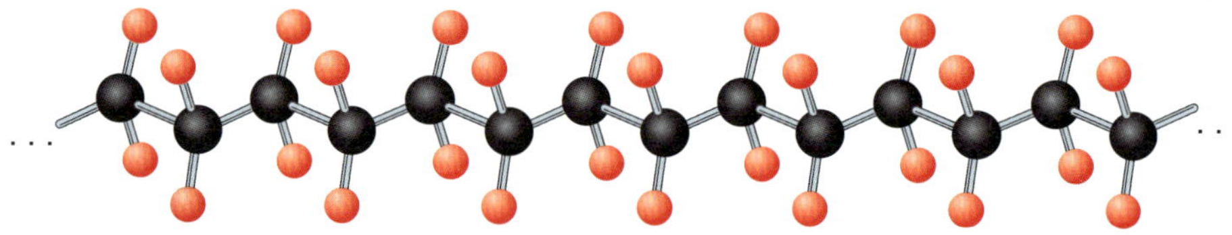

These chains or macromolecules are called **polymers**, which is another name for plastics. Polymers are composed of smaller molecules called **monomers**.

Analyse

2. What elements do we find in polymers and monomers, in addition to carbon?

1.1. The origin of plastics

Plastics can be divided into two types, according to their origin:

- **Natural plastics** are **obtained**[3] directly from vegetable or animal **sources**[4]. For example, latex and cellulose are produced by plants, and casein is a protein that is found in milk.
- **Synthetic plastics** are made with **compounds**[5] that come from petroleum, natural gas and carbon. Most plastics that we use today are synthetic.

Latex from a rubber tree

Offshore platform

Understand

3. Look at the photos and answer the questions:
   a) What is latex? Where does it come from?
   b) Find out what plastics are made from latex, cellulose and casein.
   c) What happens on offshore platforms? Who works there?

[1] **packaging**: materials, such as paper and plastic, that we use to wrap objects to protect them
[2] **molecule**: group of atoms that form a material, such as a water molecule made of hydrogen and oxygen atoms
[3] **obtain**: get something
[4] **source**: origin; where something comes from
[5] **compound**: something made from the reaction between two or more chemicals

1.2. The transformation of plastics

The industrial production of plastic from raw materials is called **polymerisation**. There are two basic steps in this process:

- During the manufacturing process, **fillers** may be added to reduce the cost of production and **enhance**[6] certain properties of the raw materials. Some typical fillers include fibreglass, paper, silica, mineral powder and sawdust.
- Chemical **additives** can also be used, such as plasticizers which increase the flexibility and resistance of the final product. Pigments may also be added to give the plastic a certain colour.

1.3. Properties of plastics

Depending on their composition, certain plastics may be more **rigid**[7], more flexible or more elastic. In general, plastics are:

- physically **resistant** to stretching, bending, twisting and compression.
- good electrical, thermal and acoustic **insulators**.
- **ductile** because they can be pulled to make threads.
- **malleable** because they can be pressed into thin sheets.
- **light** with a typical density of 0.9 to 1.3 grams per cubic centimetre (g/cc).
- **impermeable**[8] (to gas and liquids).
- **recyclable** in many ways, such as:
  - **chemical recycling**, which breaks plastic down into its chemical components.
  - **mechanical recycling**, which breaks plastic into small pieces that can be melted and made into tiny pellets.
  - **energy recovery**, which burns plastic to produce heat, electricity or energy for other industrial processes.

Non-biodegradable plastic is made from petroleum products. Most of the plastics that we use today are non-biodegradable.

Biodegradable plastic is becoming more common. Some biodegradable plastics are broken down by bacteria and other biological agents. One example is biopol, which is used to make bottles and mouldings. It can be broken down by microorganisms in soil.

Understand

5. Some polymers are silicon-based. Find information about the properties and applications of silicon polymers.
6. Are all plastics similar in hardness and elasticity? Are plastics good or bad conductors of heat and electricity?

Analyse

7. What environmental problems are caused by energy recovery?

Understand

8. Find information about bioplastics. How are they made? Why are scientists looking for new ways to produce them? What is polylactic acid (PLA)? Is it biodegradable?

Understand

4. Find information about how tyres are recycled. What processes are used? What is the result? What do we do with these materials?

[6] **enhance**: make something stronger or better
[7] **rigid**: hard and not flexible
[8] **impermeable**: not allowing things to pass through it

Key structure

We can use modal verbs with the passive.

... fillers may be added ...
Chemical additives can also be used ...

Key concepts

- Plastics consist of long chains of atoms which are mostly composed of **carbon**.
- Plastics can be classified into **natural** and **synthetic** plastics.
- The process of manufacturing plastic is called **polymerisation**.
- Plastic materials are resistant, insulating (against electricity, heat and sound), ductile, malleable, impermeable and light.
- There are three types of plastic **recycling** processes: chemical and mechanical recycling and energy recovery.

2. THE CLASSIFICATION OF PLASTICS

Analyse

9. Look at the objects in the photos. Which ones are more flexible? Which ones are more resistant? Which ones are made of the most elastic material?

2.1. Classification by internal structure

Plastic materials can be divided into three groups, according to their internal structure:

Thermoplastics are composed of polymer chains that are weakly-connected to each other.	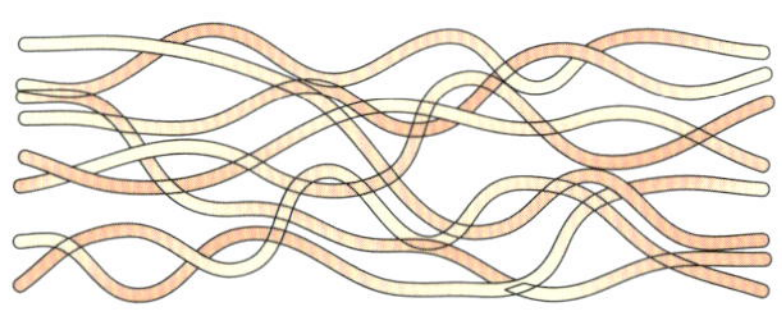	When thermoplastics are heated, they **soften**[9] and can be moulded into different shapes. The materials keep this new shape after they cool down. This heating and cooling process can be repeated as many times as needed.
Thermosetting plastics are composed of polymer chains that are strongly-connected.	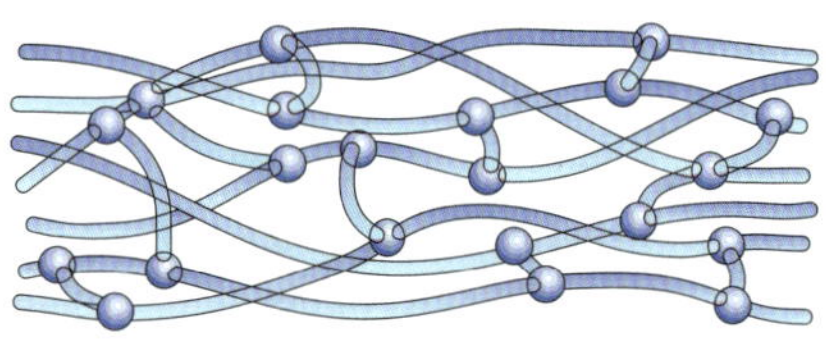	Thermosetting plastics can be heated and moulded, and they maintain that shape after they cool down. However, these materials can only be heated and shaped once. The process cannot be repeated.
Elastomers are composed of polymer chains that are laterally connected. They can be folded or rolled into a ball, like string.	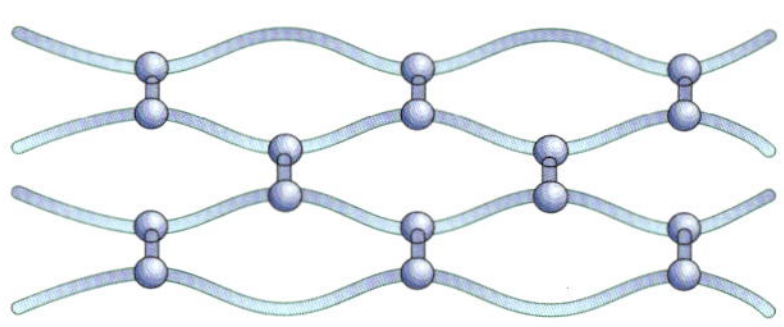	When we apply force to elastomers, they stretch because the polymer chains are very elastic. The production of elastomers includes a chemical process called vulcanisation, in which sulphur is added to rubber at 160 °C. This gives elastomers more strength, resistance and durability. However, it doesn't change their natural elasticity.

[9] **soften**: become softer

Analyse

10. Study the following objects made of plastic. Which type of plastic is probably used to make each object? Explain your answers.

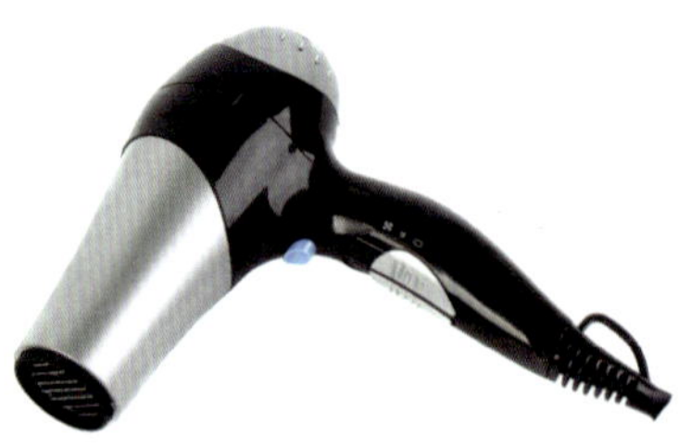

hair dryer

sponge

bottles

2.2. Industrial plastics

Many plastics are used for industrial purposes. These are classified into groups.

Thermoplastic materials			
Name		**Properties**	**Uses**
Polyethylene terephthalate 1 PET		Impermeable Transparent Light	Packaging for food, drinks, products for personal hygiene, cosmetics and cleaning products. 
High-density polyethylene 2 HDPE		Rigid Resistant	Buckets, bottles, packaging for cleaning products, medicine bottles and toys. 
Polyvinyl chloride 3 PVC		Wide range of hardness Impermeable	Windows, blinds, outer coverings of cables, water pipes, shoe soles, gloves, raincoats, garden hoses, plastic table coverings, shower curtains, umbrellas and upholstery. 
Low-density polyethylene 4 LDPE		Soft and light Transparent	Transparent film, bags, cups and plates. 
Polypropylene 5 PP		Flexible Chemical resistant Superficial hardness	Bottles, bottle caps, packaging for hygiene products and medicine, carpet fibres, bags, plates, cutlery, storage containers and garden products. 
Polystyrene 6 PS	**Moulded**	Transparent Can be coloured	CD cases, clothes hangers, cartons for eggs and dairy products, plates, cups and cutlery. 
	Expanded (Styrofoam)	Spongy Soft	Packaging, wrapping, thermal and acoustic insulation.
Methacrylate 7		Transparent Impact-resistant	Car headlights and safety glasses. 
Teflon 7		Non-stick Smooth and slippery Heat-resistant	Outer covering of airplanes, aerospace applications, non-stick cooking utensils (pots and pans), prosthetic devices and the outer covering of electrical cables. 
Cellophane 7		Transparent Flexible	Packaging for food and other products. 
Nylon 7		Highly resistant Flexible	Production of textiles tennis racket strings. 

Thermosetting plastic materials		
Name	**Properties**	**Uses**
Polyurethane (PUR)	▮ Spongy ▮ Flexible	Sponges and foam for beds or seats. 
	▮ Soft ▮ Compact	Transmission belts, friction wheels, thermal and acoustic insulators. 
	▮ Elastic ▮ Adhesive	Glue and varnish. 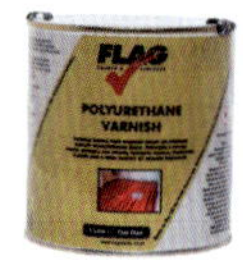
Bakelite	▮ Electrical insulator ▮ Impact-resistant when combined with fibreglass ▮ Heat-resistant when combined with asbestos	Electrical plugs and switches, pot handles, ashtrays, gears, outer casings for phones and other home appliances. 
Melamine	▮ Light, resistant and quite hard ▮ Odourless and tasteless ▮ Thermal insulator	Kitchen countertops, dishes, food containers, electrical accessories, thermal and acoustic insulators. 
Polyester resins (UP)	▮ Heat-resistant up to 200°C ▮ Strong and rigid when combined with fibreglass	Protective helmets, body parts of cars and boats, swimming pool liners, fishing rods, roofing materials. 

Understand

11. Listen to eight sentences about thermoplastic and thermosetting plastic materials. Are the sentences true or false?

Apply

12. Copy and complete the following table in your notebook. In each case, name the plastic material that you would choose. Explain the properties that it must have in order for the object to perform its function. Here's an example.

	Properties	**Material**
	spongy, flexible, elastic	polyurethane

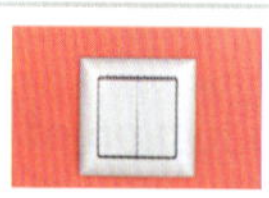		

Elastomers			
Name	**Raw material**	**Properties**	**Uses**
Natural rubber	Latex	▮ Resistant ▮ Inert	Thermal and electrical insulation, shoe soles, waterproof suits, tyres, steering wheels, bumpers, paving, pipes, hoses, sponges, gloves, mattresses.
Synthetic rubber	Petroleum products	▮ Resistant to chemical agents	
Neoprene	Synthetic rubber	▮ Enhances the properties of synthetic rubber, making it harder, more resistant and impermeable	Diving suits, joints, pipes, hoses, straps, gloves, adhesives.

Understand

13. Match the photos with the following types of plastic: methacrylate, styrofoam, nylon, PVC, polyethylene.

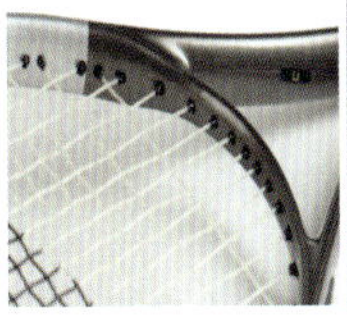

Apply

14. A company wants to make a container for storing and transporting food. What materials would you recommend? What materials would you suggest for making CD covers, car tyres or mattresses for beds?

Understand

15. What do we call the chemical process for making elastomers? What are the main steps of this process? Find information about the person who invented this process. How was it discovered?

Apply

16. Begin planning the campaign poster or leaflet for your final project. Which objects have you chosen? Can you identify the material they are made of? What are its properties? Find information about the raw materials and the process used to make the material. Use a word processing program to write informative texts.

Analyse

17. Find information about Tetra Brik cartons. What materials are used to make them? What type of plastic is used? Which function does the plastic perform?

Key concepts

- **Thermoplastics** are usually made from petroleum products. The most common thermoplastics are:
  - Polyethylene terephthalate (PET)
  - High-density polyethylene
  - Polyvinyl chloride (PVC)
  - Low-density polyethylene
  - Polypropylene
  - Moulded polystyrene
  - Expanded polystyrene or Styrofoam
- **Thermosetting** plastics are made from petroleum products. They include:
  - Polyurethane
  - Bakelite
  - Melamine
  - Polyester resins
- Typical **elastomers** include rubber and neoprene.

3. PLASTIC FORMING TECHNIQUES

Plastic products are typically made from powder, pellets or resins (thick and viscous liquids).

Various techniques can be used to form plastic products, but the most important ones are **extrusion**, **calendering**, **vacuum forming** and **moulding**.

3.1. Extrusion

The extrusion process typically includes the following steps:

1. **Thermoplastic** material, usually in the form of pellets, is added to an extrusion machine. These go into a heated cylinder where they melt and form a thick liquid.
2. Inside the cylinder, there is a screw that turns and pushes the plastic out through a die. The shape of the die gives the extruded plastic a particular form.
3. When the plastic leaves the machine, it is hardened in a bath of cool water.

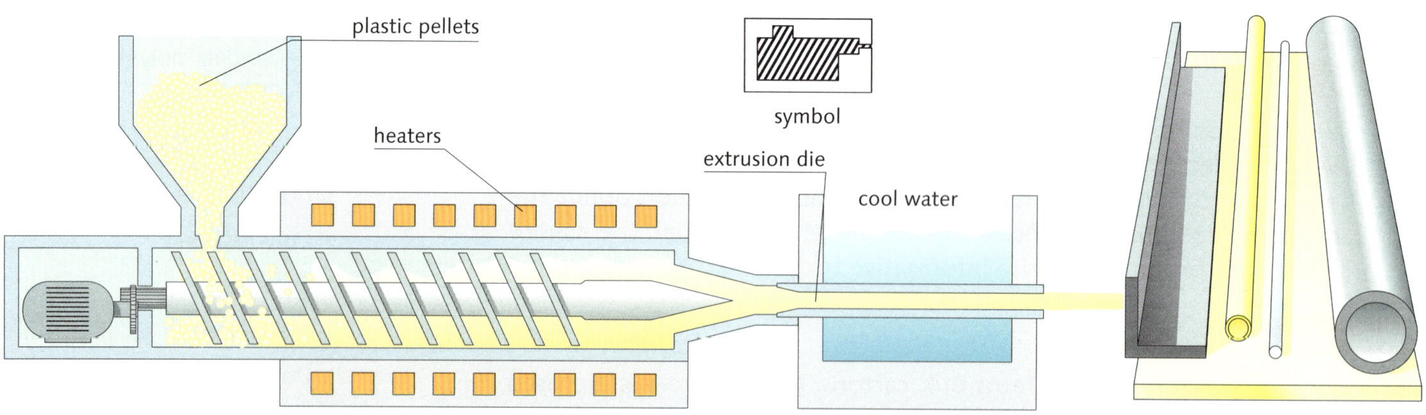

Uses: Tubes, water pipes, insulation for cables, plastic film and packaging.

3.2. Calendering

In the calendering process, **thermoplastic** material is passed between rollers to make thin sheets.

The last set of rollers ensures the plastic has the proper thickness. These rollers may also give the surface a shiny or matte finish.

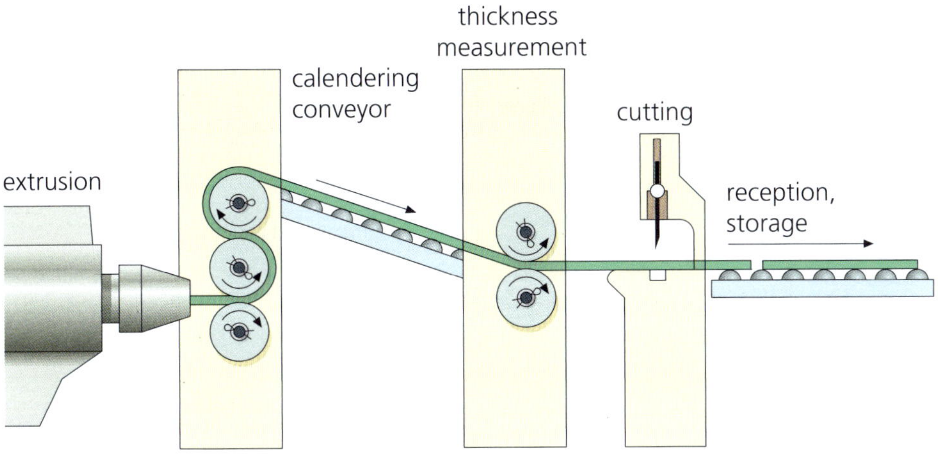

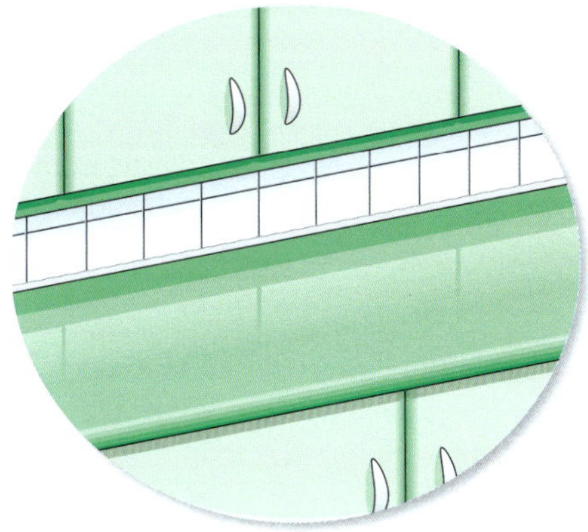

Uses: Kitchen countertops and cupboards.

3.3. Vacuum forming

The vacuum forming process involves the following steps:

1. A sheet of **thermoplastic** material is placed over a mould.
2. The sheet of plastic is heated until it becomes soft.
3. The air under the sheet is sucked out to create a vacuum. This vacuum pulls the sheet onto the mould, which gives it the desired shape.
4. The sheet is cooled and removed from the mould.

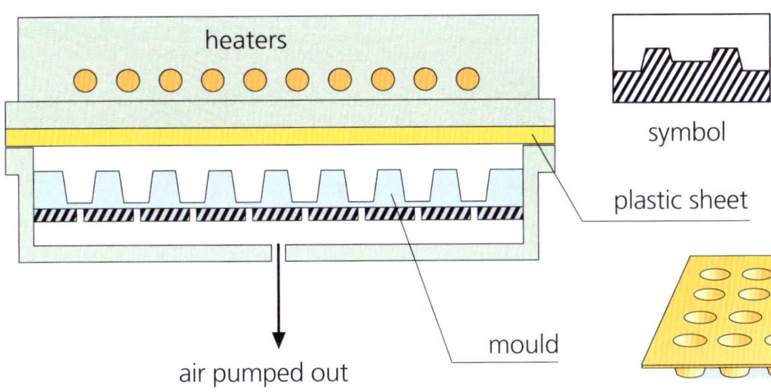

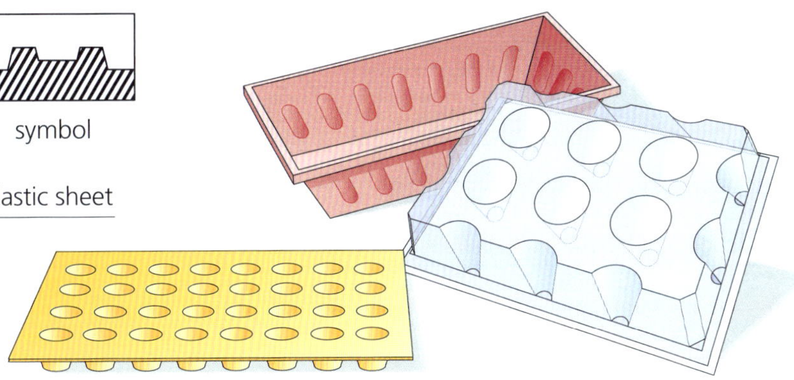

Uses: Ice trays, plastic bathtubs, car dashboards, shop signs.

Understand

18. Which techniques are used to make these products: dashboards, water pipes, kitchen countertops, bathtubs, shop signs, the covering for electric cables?

3.4. Moulding

There are various methods for using **moulds** to make plastic objects.

Blow moulding

1. A tube of **thermoplastic** material is produced by an extrusion machine. Then this tube is introduced into a hollow mould.
2. The mould is filled with compressed air, which pushes the plastic against the sides of the mould. This gives the plastic the desired shape.
3. When the plastic cools, it is removed from the mould.

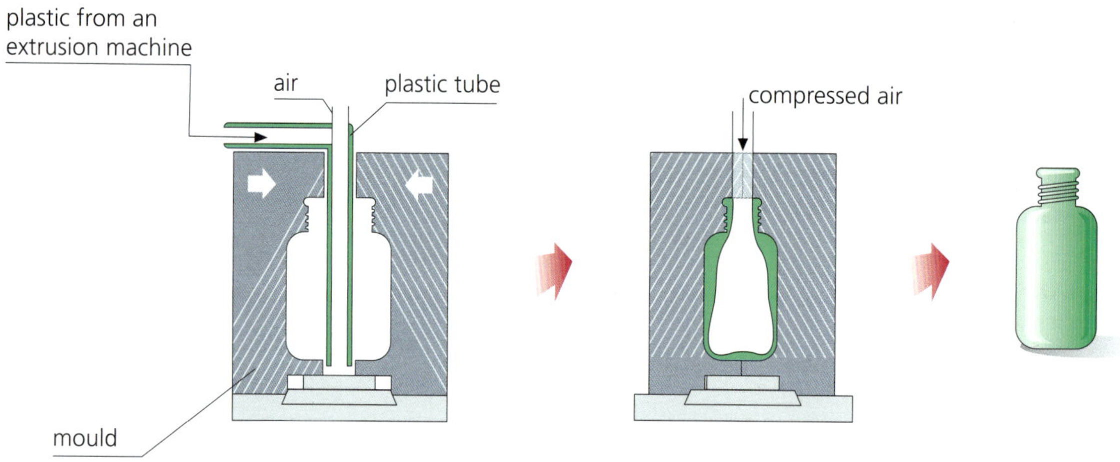

Uses: Hollow objects, such as bottles for mineral water or cooking oil; some hollow toys, such as balls.

Injection moulding

1. Melted **thermoplastic** material is introduced into a mould.
2. After the material cools down and hardens, it is removed.

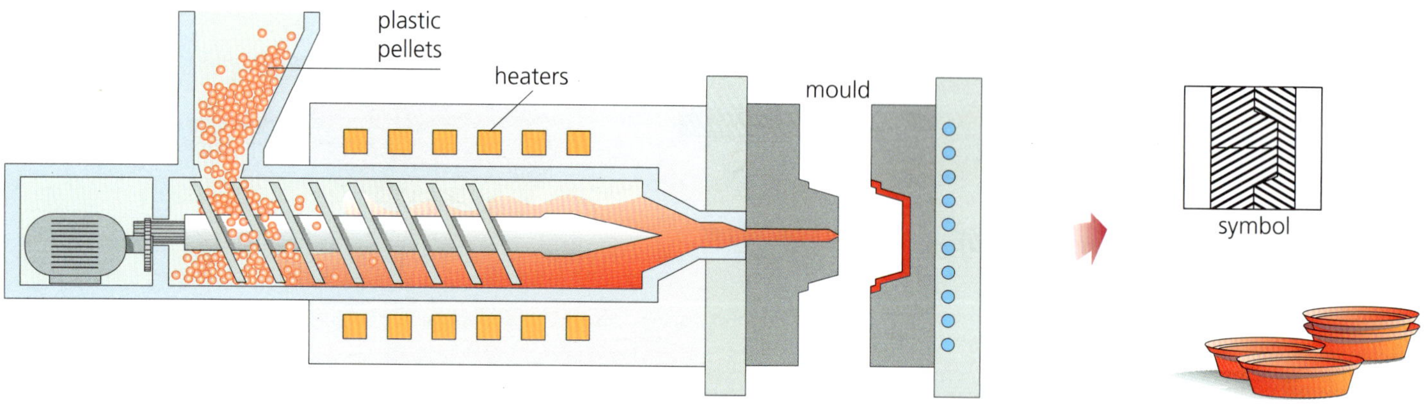

Uses: Household containers, buckets, toys and certain parts for cars, planes and spacecraft.

Compression moulding

Compression moulding is done with a machine called a press. The main steps of compression moulding are as follows:

1. **Thermosetting** plastic materials, such as pellets or powder, are introduced into a cavity mould.
2. Heating elements soften the plastic while it is compressed by the other side.
3. The plastic is shaped by the moulds on both sides.
4. After the moulds cool, the plastic object is removed.

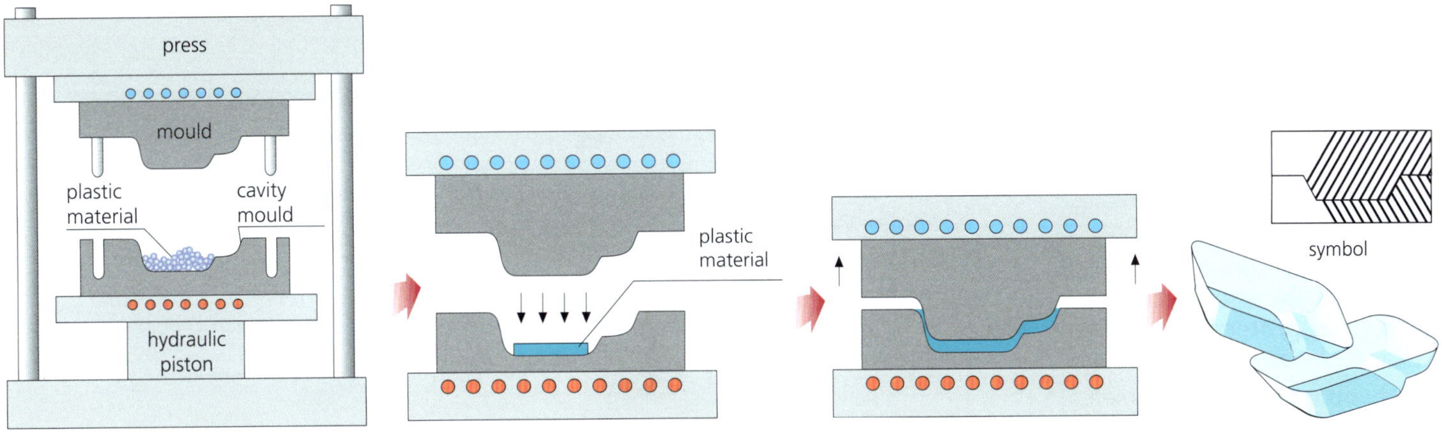

Uses: Containers and outer casings for home appliances and other machines.

Understand

19. Guess which moulding techniques are used to make the following plastic objects: a milk bottle, a baby's bathtub, the outer casing of a microwave, a child's ball.

Apply

20. Look at the photos. Which techniques are used to make these products?

Analyse

21. Find information about how Styrofoam is made. What are the properties of Styrofoam? List its main applications.

Apply

22. Think about the objects that you have chosen for your final project. What forming methods are used to produce them? Add this information to your project notes.

Key concepts

- Various industrial techniques can be used to manufacture plastic products, such as: **extrusion, calendering, vacuum forming** and **moulding**.
- The main techniques for using **moulds** are as follows: **blow moulding**, **injection moulding** and **compression moulding**.

SAFETY AND HEALTH RULES

- Use the proper tools and techniques for every task. Make sure that your tools are in perfect condition.
- Concentrate on what you are doing. Avoid distractions while you are working.
- Be careful with pointed or sharp tools. Keep your hands away from any blades, sharp edges or rotating parts.
- Keep your tools clean and tidy while you work. Put away any tools that you are not using at the moment.
- Use the proper safety equipment when necessary, such as protective goggles and gloves.
- If you cut or burn yourself, tell the teacher and get help immediately.
- Always work in an area with proper lighting and ventilation. The temperature should be comfortable and there shouldn't be distracting or irritating noises.

Understand

23. What other safety and health rules would you add to the list?

Evaluate

24. Are these rules important for workshops? Explain your opinions.

Understand

25. Find information about the following terms: health, occupational hazard, ergonomics.

4. MODIFICATION TECHNIQUES

A variety of tools and machines can be used to modify prefabricated plastic objects, such as sheets, bars and mouldings.

measuring → marking → cutting → drilling → filing and sanding → joining

4.1. Measuring

Before measuring an object, we should analyse its shape and dimensions to make the use of the material.

Tools

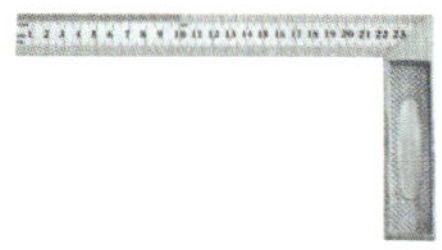

A **carpenter's square** is used to check right angles. We can use this tool to draw parallel or perpendicular lines.

A **ruler** is used for precise measurements. We can also use it to draw straight lines.

A **measuring tape** is flexible and retractable. We can use it to measure lines and surfaces that are straight or curved. If a metallic tape is wide and curved, it won't bend easily.

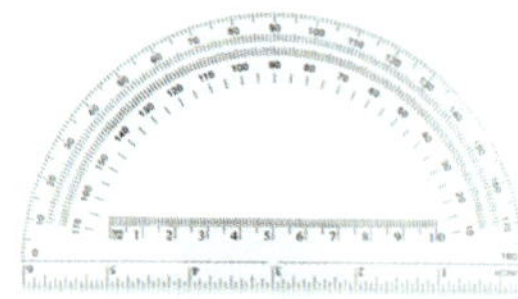

A **protractor** is a semicircular tool that we use to measure angles. Each line on a protractor represents one degree.

4.2. Drawing marks and lines

For this task you need a punch, a pencil and a permanent marker. If you don't want to use a permanent marker, you can stick a piece of paper or tape on the plastic that you are going to cut. Then you can mark the paper or tape.

Before you make any marks, think about how many pieces you are going to need and how big they need to be. Then draw a plan to use the plastic efficiently to avoid wasting material.

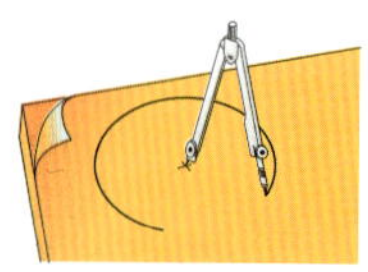

Understand

26. We use a variety of tools to hold materials in place while we are working on them. Do you recognise the tools below? When would you use them?

bench vice

C-clamp

bar clamp

ring clamp

4.3. Cutting

We can use different tools and machines for cutting plastic, depending on the hardness and thickness of the material and how we plan on cutting it.

Tools

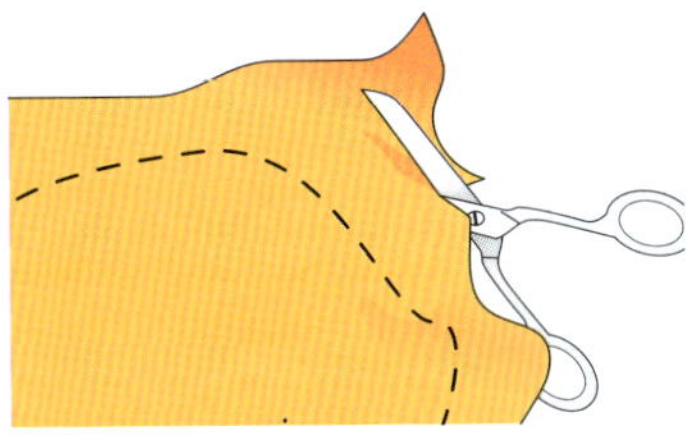

Scissors are used to cut soft, flexible sheets which are less than 1 mm thick.

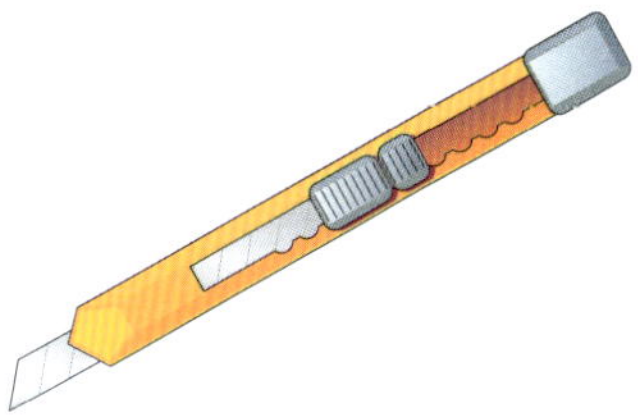

A **utility knife** can be used to cut sheets of different thicknesses depending on the hardness of the material, which may vary from 3 mm for hard PVC and PE to various centimetres for softer materials, such as Styrofoam.

When you use a utility knife, pass the blade along the cutting line several times. You can also put pressure on the line before you make the final cut.

A **hacksaw** is used to make very precise cuts. It usually has butterfly bolts to hold the blade in place. The handle may be plastic or metal.

How do you use hacksaw?

1. Draw the cutting line on the plastic material. Then use clamps or a vice to hold it in place so it doesn't move or vibrate.
2. Start by making a small cut. This makes the cutting easier.
3. While you are cutting, hold the material with your free hand.
4. Keep the hacksaw blade at a 45° angle to the material.
5. Use the full length of the blade at a moderate speed, putting more pressure when you push forward.

Machines

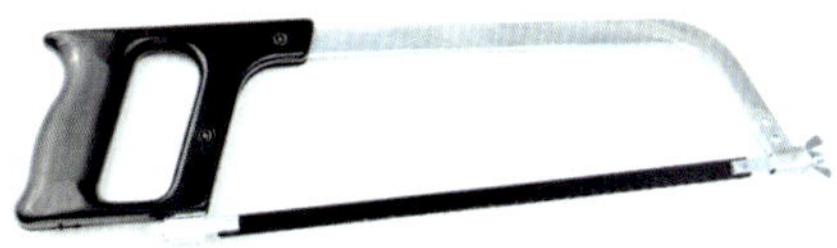

A **jigsaw** is used to cut large, rigid sheets. The blade makes a reciprocating up-and-down motion. Jigsaws can be fixed or mobile.

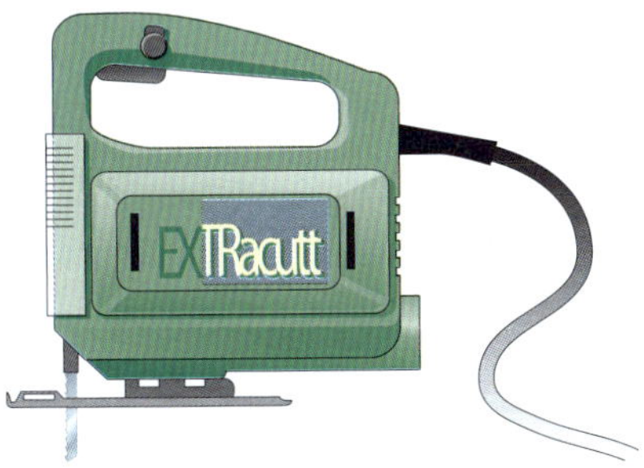

A **punch** is used to make holes in sheets that are less than 5 millimetres thick. We can use a punch to make simple shapes by applying pressure.

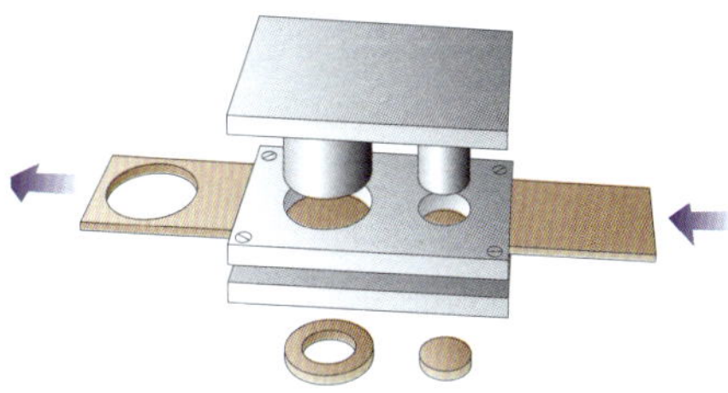

A **hot wire cutter** is used to cut thick, soft thermoplastic materials, such as polystyrene foam, or Styrofoam.

Apply

27. In your notebook, make a list of the tools or machines you would use to cut the following materials: sheets of Styrofoam, sheets of PVC that are less than 3 mm thick, large flexible sheets of plastic that are less than 1 mm thick.

4.4. Drilling

When you are drilling a piece a plastic, you must keep it from moving around.

Machines

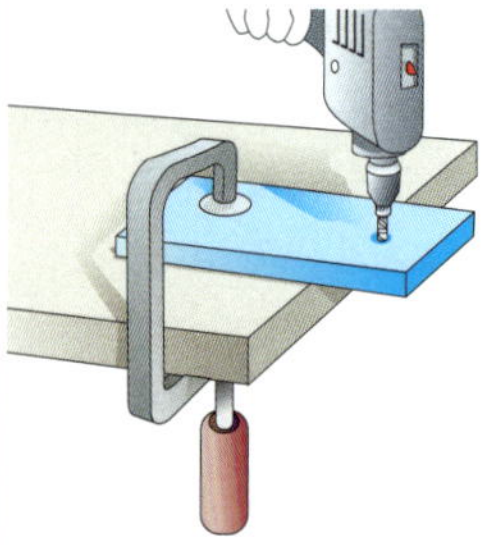

A **drill** has a metal **bit** that spins quickly and allows you to make holes in materials. A drill bit is round with a spiral groove. Bits can be made with different materials, lengths and diameters.

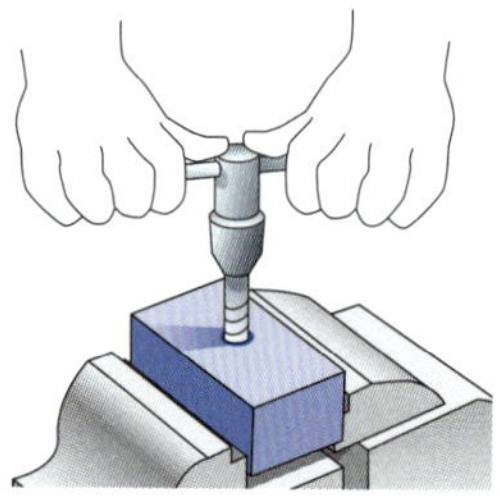

You can use a **tap and handle** to drill a piece of plastic. You should also use an appropriate lubricant while you are drilling.

Investigate

28. Find out what materials are used to make drill bits for perforating plastic.

How do you use a drill?

1. Use a vice or clamp to keep the piece of plastic in place. Then mark the place where you want to drill a hole.
2. Drill the hole carefully to avoid breaking the plastic.
3. As you drill, remove any shavings so they won't melt or get in your way.

4.5. Filing and sanding

We use various tools and methods to smooth plastic surfaces and edges.

Tools

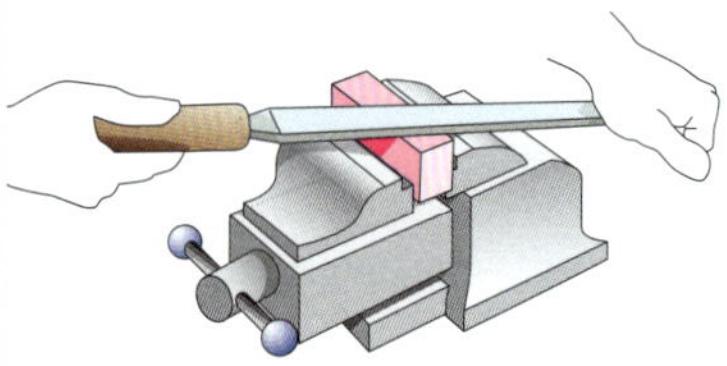

A **file** has a grooved surface that we use to remove large amounts of hard material.

You can also use **sandpaper** that is wrapped around a piece of wood.

Understand

29. Find out about the different types or grits of sandpaper. Which grits are rougher? Which ones are smoother? What materials are the made of? How are they used for different jobs?

How do you use a file?

1. Use a vice to hold the piece of plastic that you want to file. Hold the handle of the file with one hand. Hold the tip of the file with the palm of your other hand.
2. Hold the rough side of the file at a 45° angle to the piece of plastic.
3. Push down as you move the file forward. This action removes excess material.

Machines

A **belt sander** uses a continuous piece of sandpaper that is placed over two drums. The sander can be mobile or stationary.

An **orbital sander** uses a disc of sandpaper that spins very quickly.

4.6. Joining

After pieces of plastic have been shaped and modified, they can be joined together either temporarily or permanently.

Temporary joints

These joints let us connect pieces temporarily and then separate them later. They have threaded elements that avoid wear on the pieces that they join.

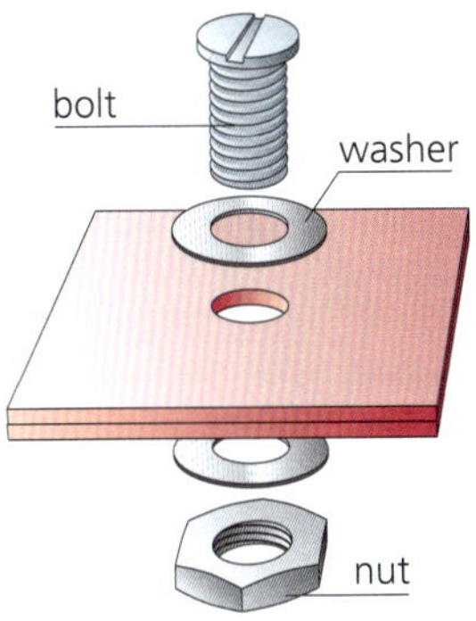

Bolts, nuts and washers are used together to form temporary joints. The bolt goes through the plastic and the nut is threaded onto the end of the bolt. Washers are placed between the plastic and the nut or the head of the bolt. This keeps the joint tight and avoids wear on the plastic.

The **threaded** pieces are joined together.

Permanent joints

We use permanent joints when we don't plan on separating the pieces later. In most cases, we cannot separate the parts without damaging or breaking them. Plastics can be joined permanently with adhesives or by soldering them.

- **Adhesives**: the choice of adhesive depends on the properties of the plastic that we are using.
  - **Two-part resins**: These adhesives are very resistant to water. The drying time is between twenty and sixty minutes.
  - **Acrylic cement**: This adhesive is used mostly for PVC and polystyrene. The drying time is approximately three hours.
  - **Contact adhesives**: These adjectives are used with all types of plastic. They join materials instantly.
- **Soldering**: Heat and pressure are used to join thermoplastic materials.

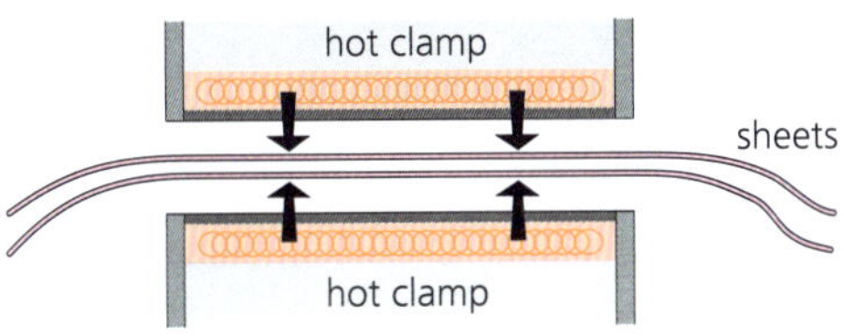

A **heat sealer** has a system of clamps that use heat and pressure to join sheets of plastic.

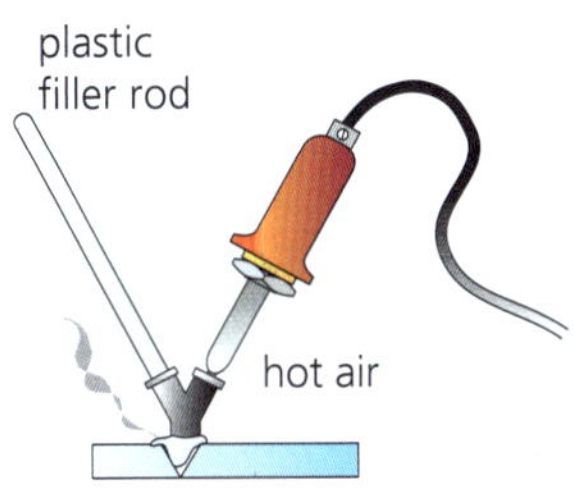

A **hot air welder** uses hot air to melt a plastic filler rod. The melted material welds pieces of bevelled plastic together.

Analyse

30. How are the following pieces joined: plastic frames for glasses, a bottle top and bottle, a plastic bag, a PVC tube, the casing of a machine, a window frame?

RULES FOR USING ADHESIVES

To join pieces of plastic correctly, follow these steps:

- Choose an appropriate adhesive and use it at the proper temperature.
- Use only as much adhesive as you really need. Make sure the pieces fit together properly and are clean and dry.
- Allow the adhesive enough time to dry properly.
- Don't get any adhesive on other areas of the pieces. Cover them to with adhesive tape keep them clean.
- Before the adhesive has set completely, use a spatula to remove any excess. Once the adhesive is dry, remove the tape.

Key concepts

- **Modification techniques** use tools and machines to make changes to prefabricated materials, such as sheets, bars or mouldings.
- Examples of these techniques include: measuring → drawing marks and lines → cutting → drilling → filing and sanding → joining.
- You must always pay special attention to health and safety rules.

5. TEXTILES

Different fibres can be used to make textiles. The fibres may be **natural** or **synthetic**.

5.1. Natural fibres

Natural fibres may come from animal, plant or mineral sources. In most cases, we clean, brush, stretch and dye the fibres, before spinning them into thread. Then we can weave the threads together to make cloth.

Understand

31. Look at photo. Complete the text with these words: *cloth*, *looms*, *threads*, *warps*, *weavers*, *wefts*. Then listen and check.

We use _____ to weave _____ . A loom has got many _____ called _____. Between those threads, the _____ pass other threads called _____ .

Natural fibres from animal sources

Wool is produced by sheep. It may be naturally white, yellowish, grey-brown or black, although wool is easily dyed other colours. Wool is very elastic and quite resistant to acids.

Silk is made by the larvae of insects called silk moths. These larvae produce a liquid that dries in contact with the air, forming thin threads. Silk is a very strong and elastic material, as well as a good thermal and electrical insulator.

Understand

32. How do sheep farmers collect wool from their sheep? What is this process called? How many kilograms of wool does a typical sheep produce every year? What other animals produce wool?

33. What do we call the larvae of silk moths? Why do they produce silk? What do they make with it? How do people collect and treat the fibres?

LEATHER

Leather is made from the hides of certain animals. The hides are cured in a series of steps which include salting, drying, soaking, shaving and scraping.

Natural fibres from plant sources

Cotton comes from the fruit of the cotton plant, which belongs to the Malvaceae family. The natural colour of cotton is white, but it can be dyed many colours. Cotton is light, elastic, permeable and resistant to acids, as well as a good thermal insulator.

Esparto comes from the leaves of esparto grass plants. The fibres are strong and resistant.

Linen comes from the flax plant, which belongs to the Linaceae family. The natural colour of linen fibres is white or beige. Linen is both elastic and flexible, as well as a good thermal conductor. Linen is also resistant to chlorine and bleach.

Bamboo comes from bamboo plants, which belong to the Graminaea family of grasses. Bamboo is porous, permeable, and biodegradable. It also has antibacterial properties.

Understand

34. What type of climate do cotton plants need to grow?

35. Which countries produce large amounts of flax?

36. What products are typically made with esparto?

37. What is bamboo used for?

Natural fibres from mineral sources

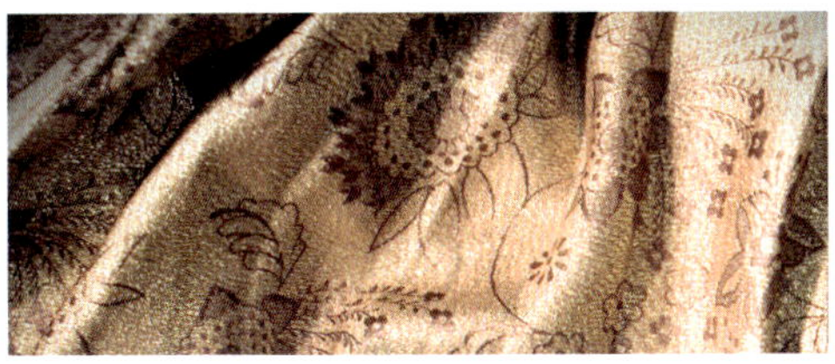

Metallic fibres can be made from gold, silver and copper. They are often used for weaving traditional costumes for religious and cultural purposes.

Analyse

38. Why are gold, silver and copper good for making long, thin fibres? What physical property have these metals got?

SELF-CLEANING TEXTILES

Some textiles, such as wool and silk, are covered with nanocrystals of titanium dioxide. When these crystals are exposed to solar radiation, they break down particles of dirt, keeping the textiles clean.

5.2. Synthetic fibres

Synthetic fibres, such as nylon, polyester, rayon and Lycra, are plastic materials. They are long-lasting, resistant and impermeable. Nowadays, a mixture of natural and synthetic fibres is often used in the production of textiles.

Apply

39. In your notebook, make a list of the natural and synthetic materials that might be used to make the following products: a waterproof coat, a sleeping bag, jeans, a summer shirt, winter boots.

40. A mixture of natural and synthetic fibres are used to make fireproof textiles. What are some uses for these textiles?

Evaluate

41. What are the advantages and disadvantages of synthetic fibres, compared to natural fibres?

Key concepts

- Both **natural** and **synthetic** fibres can be woven to make a variety of textiles.
- **Natural fibres** may come from animal sources (**wool**, **silk**), plant sources (**cotton**, **linen**, **esparto**, **bamboo**) and mineral sources (**gold**, **silver** and **copper fibres**).
- **Synthetic fibres**, such as **nylon**, **polyester**, **rayon** and **Lycra**, are plastic materials.

Constructing a vehicle

The objective is to design and build a vehicle that travels more than 1.5 m after rolling down a ramp. The ramp will have a length of 40 cm and an incline of 30°.

The vehicle must meet the following conditions:

- It must be made with recycled plastic.
- Metallic parts may not be used.

The following photos show some ideas that you may want to use.

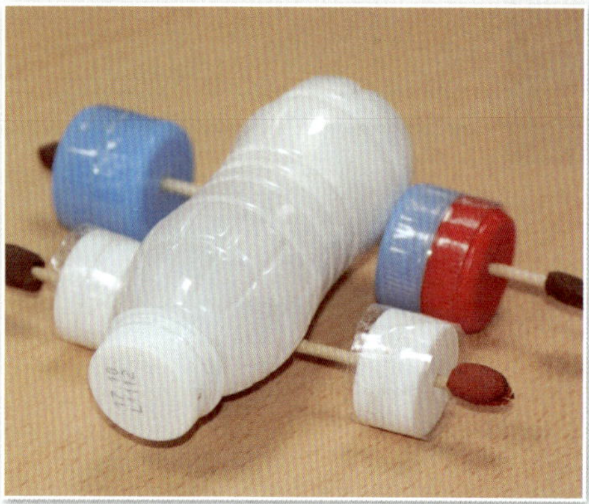

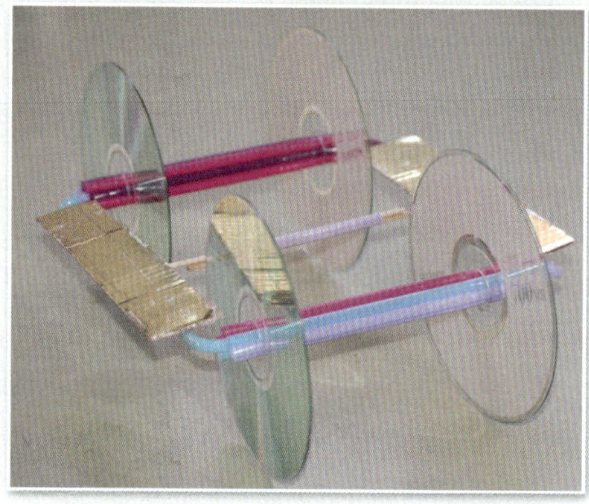

After you have finished designing your vehicle, you can begin building it. Before you start, you should revise the workshop rules for health and safety.

RECOMMENDATIONS

You should keep the following general recommendations in mind:

1. If a sheet of plastic is soft, flexible and fairly thin, you can cut it with normal scissors.
2. If you need to use a utility knife or a saw, hold the plastic sheet firmly with one hand. Then use a sharp object, such as a punch, to make a small cut or hole.
3. Be careful with the sharp edges of the plastic. Don't forget to use safety gloves. Remove any shavings or rough parts with scissors. After that, use a file.
4. Decorate your vehicle with paint or acrylic enamel.
5. If you are going to use a hot air welder, keep your hands dry.

Analyse

42. Answer the following questions:

- **a)** What materials have you used to make your vehicle?
- **b)** What processes and techniques have you used?
- **c)** What tools and machines did you need for the task?

43. Share your work with the rest of the class. Compare the different designs for the chassis and body, the wheels and the axle mechanisms. Which of your classmates' designs seem the most effective to you?

ANALYSIS OF TECHNOLOGICAL PRODUCTS

Objects made with plastic materials and textiles

The objects in the photos have been made with plastic materials and textiles that you have studied in this unit. In this section, you are going to answer questions about the products using four different methods of analysis:

- Formal analysis
- Technical analysis
- Functional analysis
- Socioeconomic analysis

FORMAL ANALYSIS

Consider the forms of the products:

1. What is the external shape (e.g., spherical, conical, prismatic)?
2. What are the external measurements (e.g., height, width, depth) in millimetres?
3. Draw each of the parts.
4. Draw the complete object.

TECHNICAL ANALYSIS

Consider the materials that have been used:

1. How many parts are there?
2. What material is each part made of?
3. How have the parts been assembled?
4. Which measurements are standardised?
5. What processes were used to make the products?

FUNCTIONAL ANALYSIS

Consider the functions of the products:

1. Why are they useful?
2. What do the various parts do?
3. How should we take care of the product?
4. What risks might there be for people who use the products?
5. What other objects could perform a similar function?

SOCIOECONOMIC ANALYSIS

Consider the social and economic effects:

1. What is the origin of the object and what needs does it satisfy?
2. In the past, how have people tried to satisfy similar needs?
3. How could the use of this product affect the environment?
4. How is the product advertised and sold?
5. How much do products like this cost?
6. Could the products be made with other materials? Would that be better? Why or why not?

44. What raw materials do we use to make plastics?

45. What does the word 'polymerisation' mean?

46. What happens to products that are recycled mechanically?

47. What are the differences between moulded polystyrene and Styrofoam?

48. Why are thermosetting plastics used to make objects that will be exposed to heat?

49. What thermosetting plastics are used to make acoustic insulation?

50. Why is fibreglass added to polyester resins?

51. Where does natural rubber come from? How is synthetic rubber made? List some applications for these elastomers.

52. Three plastics (A, B and C) have the following structures:

A: Long chains of molecules with weak connections between them.

B: Long chains of molecules with lateral connections between them.

C: Long chains of molecules with strong connections between them.

a) In your notebook, draw a diagram for each type of structure.

b) Label the diagrams with the following terms: thermoplastic, thermosetting plastic, elastomer.

c) Explain your answers.

53. In your notebook, match the following thermosetting plastic materials with the applications below: melamine, polyurethane, Bakelite, polyester resin.

a) the hull of a boat

b) a kitchen countertop

c) the handle of a frying pan

d) acoustic insulation

54. Describe the process of vacuum forming.

55. In your notebook, make a list of the tools and machines that are used for cutting, perforating and smoothing plastic materials.

56. Think about the following sentence: 'Some plastics have a memory.' What do you think this means?

57. In your notebook, make a list of the types of plastic that are used in the following sectors: medicine, transport, construction, textile manufacturing.

58. Look at the photos below. In your notebook, make a list of the types of plastic that are used to make each object. Use books or the Internet to find information. Then classify the materials as thermoplastics, thermosetting plastics or elastomers.

59. Answer the following questions in your notebook:

a) What plastics can be heated and reshaped many times?

b) What plastic is used to make transparent film for wrapping food?

c) What material is used to make the outer casing of home appliances?

d) What process is used to make plastic tubes and pipes?

e) What forming techniques are used to make plastic bottles?

f) What tool is used to cut Styrofoam?

60. Below is a photo of celluloid. Find information about this material. How was it invented? What was it used for? Why is it considered one of the first plastics?

61. Which plastics are used as fibres to make textiles? Look at the tags of your clothes for ideas.

62. Find information about the properties and uses of Kevlar and Nomex.

63. Plastics have allowed scientists to make great medical advances. For example, we now use plastic to make heart valves and other artificial body parts. Find information about how plastics are used for medical purposes.

STUDY SKILLS

- Write a summary using the key concept boxes in the unit. Include everything you think is important.
- Draw a mind map of the concepts in the unit. Use the following structure to get started:

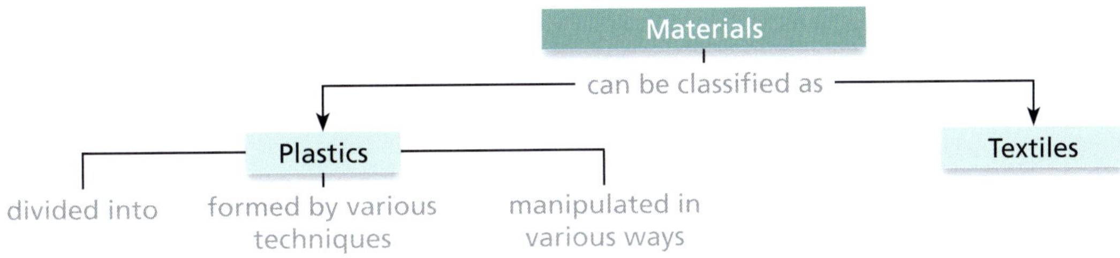

- Make a technical dictionary with the terms that you think are the most important. Include: polymerisation, thermoplastics, thermosetting plastics, elastomers, extrusion, and calendering. Then add other terms that you think are important.

FINISHING THE FINAL TASK

Designing an information campaign about plastics

Once you have finished your poster or leaflet, consider the following suggestions:

1. Before you begin your campaign:
   - Show your poster or leaflet to your classmates, explaining each of the sections that were listed in the presentation of the project.
   - If you have designed a survey, talk about the information that you have collected and the most interesting details, such as the average age of the people that you interviewed, how much they knew about plastics, the most commonly used plastics or how many of them recycle.
   - Discuss how you are going to do the campaign. Talk about where, when and how long it will take place. How will you get people's attention?

2. When you have finished your campaign:
   - Exchange opinions about the experience. How did people react to your campaign? How interested were they? Are you satisfied with the results?

OXFORD
UNIVERSITY PRESS

Oxford University Press is a department of the University of Oxford. It furthers the University's objective of excellence in research, scholarship, and education by publishing worldwide. Oxford is a registered trade mark of Oxford University Press in the UK and in certain other countries.

Published in Spain by
Oxford University Press España S. A.
Parque Empresarial San Fernando, Edificio Atenas
28830 San Fernando de Henares, Madrid, Spain

ISBN: 978-84-673-9366-8
D.L.: M-26439-2015
Printed in Spain

AUTHOR OF THE ADAPTED TEXT

Robert Quinn

AUTHORS

Jesús Moreno Márquez
M.ª Victoria Salazar Nicolás
Araceli Isabel Sánchez Sánchez
Francisco Javier Sepúlveda Irala

DESIGN

Interiors: Dolors Albareda

ILLUSTRATION

Luis Bogajo Peñarroya, Guillermo Díez Celaya

PHOTOGRAPHS

ACI, AGE Fotostock, Alamy Images, Lucia Gallo, Jesús G. Reyes, Getty Images, Iñaki Preysler, Jorge Montoro, Prisma, M.ª Victoria Salazar Nicolás, Araceli Isabel Sánchez Sánchez, Shutterstock (p. 4: Mike Flippo, Holbox; p. 6: Didecs, Kumma, Christopher Poe, K. Miri Photography; p. 8: Nenov Brothers Images, GeorgeMPhotography, Crepesoles, Nordling, Taweesak Thiprod, Igor Sh.; p. 9: África Studio, Maxx-Studio, Pavel Kubarkov, Coprid, Evgeny Karandaev, Alzay, Petr Malyshev, AlenKadr; p. 10: Pshenichka, Alex459, Ayazad, Jurijs Korjakins, Keerati, Galyna Nabatnikova, África Studio, Photo Love; p. 11: K. Miri Photography, Neamov, Sima, Tankist276, Coprid, Roman Samokhin, Rtimages; p. 12: Rangizzz, Coprid, Underverse, Samoshkin, Casadaphoto, Sorapop Udomsri; p. 15: Fotofermer, Axel Bueckert, Vitalina Rybakova, NanD_Phanuwat, Everything; p. 16: Sevenke, Windu, Didden, Robert_s, Cherezoff, Ason Salmon, Vseb, Mastio1; p. 17: Phrej; p. 18: Norman Pogson, VladaKela; p. 20: María Dryfhout, Jerry Horbert; p. 21: Windu, Israel Hervas Bengochea, BMCL, Sitti; p. 23: Winai Tepsuttinun, Kedrov; p. 24: Buchaka Alexander, Xpixel, DeanHarty, AlinaMD), and the Oxford Archive